The Invisible Power

Harness the Energy of Attraction to Create the
Reality You Desire

by
Joseph La Belle

Introduction

- Chapter 1: The power of attraction: what it is and how it works
- Chapter 2: Unmasking Limiting Beliefs that Prevent You from Attracting What You Want
- Chapter 3: How to Set Clear and Powerful Goals Using the Power of Attraction

Part 1: Aligning Your Thoughts and Emotions

- Chapter 4: The Power of Gratitude: How to Attract More of What You Appreciate
- Chapter 5: Creative Visualisation: Turning Your Dreams into Reality
- Chapter 6: Positive Affirmations: Programming Your Mind for Success
- Chapter 7: Releasing Negative Emotions: Freeing Yourself from the Baggage of the Past
- Chapter 8: Practising Forgiveness: Freeing Yourself and Others to Attract the Best
- Chapter 9: Raising Your Energetic Vibration: Attracting Positive Experiences

The power of attraction could work through a similar mechanism, in which our positive expectations influence our immune system, our hormones and our biochemistry, creating the conditions to attract what we desire.

- **Neuroplasticity**: The brain is a plastic organ that constantly changes and adapts according to our experiences and thoughts. By focusing on what you want, you would create new neural patterns that would increase the likelihood of attracting those experiences into your life.

Regardless of the exact mechanism, the power of attraction is a powerful tool that can be used to create a happier and more fulfilling life.

Examples of how to apply the power of attraction in daily life

- Set clear and specific goals. What do you really want to achieve in your life? Write down your goals in a detailed and positive way.

- Visualise your desires. Spend time every day to imagine that you have already achieved your goals. Use all your senses to make the visualisation as real as possible.

- Affirm what you want. Repeat positive sentences about yourself and your goals. This will help you programme your subconscious mind for success.

- Be grateful for what you have. Gratitude is a powerful emotion that can attract even more positive things into your life. Take time each day to express gratitude for the good things around you.

- Help others. Helping others is an excellent way to raise your energetic vibration and attract more positivity into your life.

The power of attraction is not a magic wand, but it is a powerful tool that can be used to create a better life. If you are willing to put in the time and effort to apply its principles, you can begin to attract more happiness, success and abundance into your life.

Remember, the power of attraction is always at work, whether you are aware of it or not. The key is to align your thoughts, emotions and actions with what you desire and allow the universe to work its magic.

**In this chapter we have only introduced the concept of the power of attraction. In the next chapters we will explore in a more practical way how to apply it.*

Chapter 2: Unmasking limiting beliefs that prevent you from attracting what you want

Introduction

The power of attraction is a powerful principle that can be used to create a happier and more fulfilling life. However, many people fail to achieve the desired results because they are blocked by limiting beliefs.

Limiting beliefs are negative beliefs we have about ourselves, the world and what is possible. They are often based on past experiences, family patterns or social influences. These beliefs can be very powerful and can unconsciously sabotage our efforts to attract what we desire.

How to identify limiting beliefs

The first step in overcoming limiting beliefs is to recognise them. Here are some signs that might indicate the presence of negative beliefs:

- You feel that you are not good enough. Do you constantly compare yourself to others and feel inferior?

- You are afraid of failure. Do you avoid taking risks or trying new things for fear of failure?

- You do not believe you deserve what you desire. Do you think that success or happiness is reserved for only a few lucky people?

- You unconsciously sabotage yourself. Do you make choices that take you away from what you desire, even though you know they are not the right thing to do?

If you recognise yourself in any of these signs, it is likely that you have limiting beliefs that are preventing you from attracting what you desire.

How to overcome limiting beliefs

Once you have identified your limiting beliefs, you can start to overcome them. Here are some suggestions:

- Question your beliefs. Ask yourself whether your limiting beliefs are based on facts or on fears and insecurities.

- Reformulate your beliefs. Replace your limiting beliefs with positive and powerful affirmations.

- Visualise success. Imagine that you have already achieved your goals and feel happy and fulfilled.

- Surround yourself with positive people. Spending time with people who support and encourage you can help you maintain a positive mindset.

- Celebrate your successes. Every accomplishment, no matter how small, is a step towards achieving your goals.

Overcoming limiting beliefs takes time and commitment, but it is possible. With the right mindset and the right strategies, you can break free from the negative beliefs that are holding you back and start attracting what you want in your life.

Exercise: Identify your limiting beliefs

Take some time to reflect on your beliefs about yourself, the world and what is possible. Write down a list of all the negative beliefs that come to mind.

Once you have identified your limiting beliefs, you can start to question them and replace them with positive affirmations. This exercise can be an important first step in overcoming your negative beliefs and attracting more happiness and success into your life.

Remember, you are more powerful than you think. You have the power to create the life you desire. Start unmasking your limiting beliefs and start attracting what you deserve.

Chapter 3: How to set clear and powerful goals using the power of attraction

Introduction

The first step to attracting what you want in your life is to set clear and powerful goals. Clear goals give you direction and help you focus on what is really important to you. Powerful goals motivate you and push you to take action.

How to set clear goals

A clear goal is one that is specific, measurable, achievable, relevant and with a defined deadline (SMART). Here are some tips for setting clear goals:

- Choose a goal that you are passionate about. If you are not passionate about your goal, it will be difficult to find the motivation to achieve it.

- Be specific. The more specific your goal is, the better you will know what you need to do to achieve it.

- Make your goal measurable. This way you can monitor your progress and see how close you are to reaching your goal.

- Make sure your goal is achievable. Do not set a goal that is too difficult or unattainable, otherwise you will become discouraged.

- Check that your goal is relevant to your life. Your goal should be something that really interests you and that will help you achieve your long-term goals.

- Give your goal a defined deadline. This will help you create a sense of urgency and stay focused.

How to make your goals powerful

A powerful goal is one that motivates you and pushes you to take action. Here are some tips for making your goals powerful:

- Visualise achieving your goal. Imagine how you will feel when you have achieved your goal. This will help you stay motivated.

- Affirm your goal. Repeat your goal to yourself every day. This will help you programme your subconscious mind for success.

- Break your goal down into small steps. A big goal can seem daunting. By breaking it down into smaller, more manageable steps, you will make it more attainable.

- Celebrate your successes. Every achievement, no matter how small, is a step towards reaching your goal.

- Find a support system. Surround yourself with people who believe in you and will help you reach your goals.

Using the power of attraction to reach your goals

The power of attraction can be a powerful tool to help you reach your goals. Here are some tips for using the power of attraction to set and achieve your goals:

- Write down your goals. This will help you clarify your goals and make them more real.

- Create a vision board. A vision board is a collection of pictures and words that represent your goals. Creating one can help you visualise the achievement of your goals and stay motivated.

- Practice gratitude. Gratitude is a powerful emotion that can help you attract more positive things into your life. Take time every day to express gratitude for what you already have.

- Affirm your goal. Repeat your goal to yourself every day with conviction and positivity.

- Visualise the achievement of your goal. Imagine how you will feel when you have achieved your goal. This will help you programme your subconscious mind for success.

- Take action. Don't wait for things to happen on their own. Take action to achieve your goals.

The power of attraction is a powerful tool that can help you achieve your goals. If you are willing to put in the time and effort to apply its principles, you can begin to attract more happiness, success and abundance into your life.

Remember, you are capable of achieving great things. Set clear and powerful goals, use the power of attraction and never give up. With hard work and dedication, you can achieve anything you set your mind to.

Chapter 4: The power of gratitude: how to attract more of what you appreciate

Introduction

Gratitude is one of the most powerful emotions we can experience. It is the ability to appreciate the good things in our lives, big or small. When we practise gratitude, we focus on the positive and this helps create a higher energetic vibration that attracts even more positive things into our lives.

How gratitude works

Gratitude works in several ways to attract what we appreciate:

- It raises our energetic vibration. When we are grateful, we feel happy and positive. This high energy vibration attracts positive experiences and people to us.

- It helps us focus on what is important. When we focus on the things we are grateful for, we stop focusing on the negative things in our lives. This helps us attract more of what we want.

- It opens our minds to new possibilities. When we are grateful, we are more open to new experiences and opportunities. This can help us achieve our goals and create a better life.

How to practise gratitude

There are many ways to practice gratitude. Here are some suggestions:

- Keep a gratitude diary. Every day, write down three things you are grateful for. They can be big things, like your health or your family, or small things, like a good day or a cup of delicious coffee.

- Practice gratitude throughout the day. Take a moment each day to stop and appreciate the beautiful things in your life. You can do this by noticing the beauty of nature, thanking someone for their kindness or simply taking a moment to breathe deeply and appreciate the present moment.

- Create a ritual of gratitude. You can create a daily or weekly ritual to practise gratitude. This could include keeping a gratitude journal, praying or meditating, or simply spending time reflecting on the things you are grateful for.

- Talk about your gratitude. Share the things you are grateful for with your friends, family and colleagues. This can help spread positivity and inspire others to practice gratitude in their lives.

Exercise: Practise gratitude for a week

Choose one week to focus on gratitude. Each day, write down three things you are grateful for in your journal. Also take time each day to stop and appreciate the good things in your life. At the end of the week, reflect on how you are feeling. Have you noticed any changes in your mood or perspective?

Gratitude is a powerful tool that can help you create a happier and more fulfilling life. Start practising gratitude today and see how your life changes for the better.

Remember, gratitude is not just something you do, it is a way of life. When you are grateful for the good things in your life, you attract even more things to appreciate.

Chapter 5: Creative visualization: turning your dreams into reality

Introduction

Creative visualization is a powerful technique that can be used to turn your dreams into reality. Through visualization, you can create vivid and realistic images of what you want to achieve in your life. This can help you program your subconscious mind for success and attract the things you want.

How creative visualization works

Creative visualization works in several ways:

- Program your subconscious mind. Your subconscious mind is responsible for 95 percent of your thoughts and actions. When you visualize your goals, you are programming your subconscious mind to believe they are possible and to help you achieve them.

- It increases your motivation. When you can see your goals clearly, you are more motivated to take action to achieve them.

- Reduces stress and anxiety. Visualization can help you relax and focus on the positive. This can reduce stress and anxiety, which can hinder your success.

- It boosts your self-confidence. When you visualize yourself achieving your goals, it increases your self-confidence. This makes you more willing to take risks and seize opportunities as they arise.

How to practice creative visualization

Here are some tips for practicing creative visualization:

- Find a quiet place where you will not be disturbed.

- Close your eyes and relax.

- Breathe deeply for a few minutes.

- Imagine that you have already achieved your goal.

- Create a vivid, detailed image of what you desire.

- Use all your senses to make the visualization as real as possible.

- Feel the positive emotions you would feel if you reached your goal.

- Repeat the visualization every day.

Exercise: Visualize achieving your goal

Choose a goal you wish to achieve and devote 10 minutes each day to visualizing it. Follow the steps above to create a vivid, detailed image of yourself achieving your goal. Feel the positive emotions you would feel if you achieved it. Repeat the visualization every day for a week and then notice how you feel. Have you noticed any changes in your mood or motivation?

Creative visualization is a powerful tool that can help you achieve your goals. If you are willing to put in the time and effort to practice it, you can begin to turn your dreams into reality.

Remember, visualization is not just a mental exercise; it is a way of life. When you consistently visualize your goals, you are sending a powerful message to the universe that you are ready to receive them.

Chapter 6: Positive affirmations: programming your mind for success

Introduction

Positive affirmations are short, powerful phrases that you can repeat to program your mind for success. When you make positive affirmations, you are essentially telling your subconscious mind what you want to believe about yourself and your world. This can have a profound impact on your life, helping you achieve your goals and create a happier, more fulfilling life.

How positive affirmations work

Positive affirmations work in several ways:

- They remove limiting beliefs. Limiting beliefs are negative beliefs we hold about ourselves and the world. They can hinder our success by preventing us from believing that we are capable of achieving our goals. Positive affirmations can help replace these limiting beliefs with positive, reinforcing beliefs.

- They boost your self-esteem. Self-esteem is your belief in your worth and abilities. When you make positive affirmations about yourself, you increase your self-esteem. This makes you more willing to take risks and seize opportunities as they arise.

- It focuses your attention on what is important. When you make positive affirmations, you are focusing your attention on what you want to achieve in your life. This can help you attract more positive things into your life.

- They reduce stress and anxiety. Positive affirmations can help you relax and focus on the positive. This can reduce stress and anxiety, which can hinder your success.

How to make positive affirmations

Here are some tips for making positive affirmations:

- Choose affirmations that are positive and specific. Your affirmations should be positively worded and should focus on what you want, not what you don't want. They should also be specific and targeted to your specific goals.

- Repeat your affirmations with conviction. When you make your affirmations, you should repeat them with conviction and faith. You should believe that your affirmations are true and can help you achieve your goals.

- Make your affirmations regularly. Ideally, you should make your affirmations several times a day, such as in the morning and in the evening. You can also do them whenever you feel down or need a boost of motivation.

- Visualize the achievement of your goals. When you make your affirmations, it is also helpful to visualize the achievement of your goals. This can help you program your subconscious mind for success.

- Act on your goals. Positive affirmations alone are not enough to achieve your goals. You must also take action to achieve your goals. However, positive affirmations can help you stay motivated and focused on your goals.

Exercise: Create your positive affirmations

Think of a goal you wish to achieve. Create 5 positive affirmations that are specific, focused and positively worded. Repeat your affirmations with conviction every day for a week and then notice how you feel. Have you noticed any change in your mood or motivation?

Positive affirmations are a powerful tool that can help you achieve your goals and create a happier, more fulfilling life. If you are willing to put in the time and effort to practice them, you can begin to see a positive change in your life.

Remember, positive affirmations are not a magic spell, but a powerful tool that can help you program your mind for success. When you make positive affirmations with conviction and faith, you are sending a powerful message to the universe that you are ready to receive what you desire.

Chapter 7: Releasing negative emotions: getting rid of the baggage of the past

Introduction

Negative emotions, such as anger, resentment, guilt and shame, can have a significant impact on our lives. They can hinder our happiness, success and relationships. If we want to attract what we desire into our lives, it is important to learn how to release negative emotions and shed the baggage of the past.

Why it is important to release negative emotions

Negative emotions can have a negative impact on our lives in several ways:

- They prevent us from being happy. When we focus on negative emotions, it is difficult to feel happy and fulfilled.

- They hinder our success. Negative emotions can negatively influence our decisions and actions, preventing us from achieving our goals.

- They damage our relationships. Negative emotions can lead to conflict and problems in our relationships with others.

- They affect our physical health. The stress and anxiety associated with negative emotions can have a negative impact on our physical health.

How to release negative emotions

There are many ways to release negative emotions. Here are some suggestions:

- Acknowledge your emotions. The first step in releasing negative emotions is to acknowledge them. Take time to reflect on how you are feeling and identify the emotions that are causing you problems.

- Accept your emotions. It is important to accept your emotions, both positive and negative. Do not try to suppress or deny them.

- Express your emotions. Find a healthy way to express your emotions. This might involve talking to a friend or therapist, writing a journal, or practicing relaxation techniques such as meditation or yoga.

- Forgive yourself. If you are struggling with feelings of guilt or shame, it is important to forgive yourself. Everyone makes mistakes, and forgiveness is essential to moving forward.

- Let go of the past. You cannot change the past, but you can choose not to let it control you. Let go of the negative experiences of the past and focus on the present and the future.

Exercise: Release your negative emotions.

Choose a negative emotion that you are struggling with. Find a quiet place where you will not be disturbed. Close your eyes and breathe deeply for a few minutes. Imagine releasing the emotion from your body. You can visualize the emotion leaving your body in the form of smoke, light or whatever else seems appropriate. Continue to breathe deeply and focus on releasing the emotion. When you feel ready, open your eyes and bring your attention back to the present.

Releasing negative emotions is not easy, but it is an important process in creating a happier and more fulfilling life. If you are willing to put in the time and effort to work on your negative emotions, you can begin to shed the baggage of the past and attract more positive things into your life.

Remember, you are not alone. Everyone struggles with negative emotions from time to time. There are many resources available to help you overcome your negative emotions and create a happier, healthier life.

Chapter 8: Practicing forgiveness: freeing yourself and others to attract the best

Introduction

Forgiveness is one of the most powerful tools we have for attracting the best into our lives. When we forgive someone, we are not saying that we approve of their actions. We are simply choosing to get rid of the resentment, anger and other negative emotions that are holding us back. Forgiveness is an act of love and compassion, both for ourselves and for the other person.

Why forgiveness is important

Forgiveness is important for several reasons:

- It frees us from pain. When we hold on to resentment and anger, we are essentially poisoning ourselves. Forgiveness allows us to get rid of these negative emotions and experience peace and happiness.

- It improves our relationships. When we forgive someone, we open the door to reconciliation and healing. This can improve our relationships with others and create a more loving and connected life.

- It allows us to move forward. The past cannot be changed, but we can choose not to let it control us. Forgiveness allows us to let go of negative experiences from the past and focus on the present and the future.

- It increases our self-esteem. When we forgive others, we are essentially proving to ourselves that we are worthy of love and happiness. This can increase our self-esteem and self-confidence.

How to practice forgiveness

Practicing forgiveness is not always easy, but it is an important process in creating a happier and more fulfilling life. Here are some suggestions:

- Recognize that you need to forgive. The first step to forgiveness is to recognize that you need to do it. This can be difficult, especially if you have been hurt by someone. But it is important to remember that forgiveness is not about the other person, it is about yourself and your peace of mind.

The Invisible Power

- Choose to forgive. Once you recognize that you need to forgive, you must choose to do so. This does not mean that you have to forget what happened, but it does mean that you have to get rid of the resentment and anger you are feeling.

- Accept that forgiveness is a process. Forgiveness is not something that happens overnight. It takes time and commitment. Be patient with yourself and don't give up.

- Find a way to express your forgiveness. There are many ways to express forgiveness. You can talk to the person who hurt you, write a letter of forgiveness, or simply forgive them silently in your heart.

- Let go. Once you have forgiven someone, it is important to let go. This means to stop thinking about the past and focus on the present and the future.

Exercise: Practice forgiveness

Think about someone who has hurt you. Take some time to reflect on how you feel about this person. Are you ready to forgive them? If yes, how can you express your forgiveness? If no, what prevents you from forgiving her?

Forgiveness is a gift we give to ourselves. When we forgive someone, we free ourselves from pain and suffering and open the door to a happier and more fulfilling life. If you are willing to put in the time and effort to practice forgiveness, you can begin to live a freer, more loving life.

Remember, forgiveness is not a sign of weakness; it is a sign of strength. It is a choice to free yourself from the past and create a better future for yourself.

Chapter 9: Raising your energetic vibration: attracting positive experiences

Introduction

Everything in the universe is energy, including our bodies and thoughts. A person's energy vibration is the frequency at which his or her energy vibrates. A high energy vibration is associated with positive emotions such as happiness, love and gratitude. A low energy vibration is associated with negative emotions such as anger, fear and sadness.

How energy vibration works

Our energy vibration affects our lives in many ways:

- It attracts similar experiences. The law of attraction says that what is similar attracts what is similar. So if you have a high energy vibration, you will attract positive experiences into your life. If you have a low energy vibration, you will attract negative experiences.

- It affects our physical and mental health. A high energy vibration is associated with better physical and mental health. A low energy vibration is associated with worse physical and mental health.

- It affects our relationships. A high energy vibration is associated with more positive and fulfilling relationships. A low energy vibration is associated with more negative and conflicting relationships.

How to raise your energy vibration

There are many things you can do to raise your energetic vibration:

- Focus on positive thoughts and emotions. Your thoughts and emotions have a powerful impact on your energy vibration. Focus on positive thoughts and emotions as often as possible.

- Practice gratitude. Gratitude is a powerful way to raise your energetic vibration. Take time each day to express gratitude for the good things in your life.

- Spend time in nature. Nature has a very high energetic vibration. Spending time in nature can help you raise your energetic vibration.

- Meditate and practice yoga. Meditation and yoga can help you calm your mind and raise your energy vibration.

- Surround yourself with positive people. The people you spend time with affect your energy vibration. Surround yourself with positive and supportive people.

- Avoid negative people and situations. Negative people and situations can lower your energy vibration. Try to avoid them as much as possible.

Exercise: Elevate your energetic vibration.

For the next seven days, focus on positive thoughts and emotions. Practice gratitude every day. Spend time in nature. Meditate or practice yoga for a few minutes each day. Surround yourself with positive people. Avoid negative people and situations. At the end of the week, notice how you feel. Have you noticed any changes in your mood or energy?

Elevating your energy vibration is one of the most powerful ways to improve your life. When you have a high energy vibration, you will attract more positive experiences, have better physical and mental health, and have more positive relationships. If you are willing to put in the time and effort to raise your energy vibration, you can begin to create a happier and more fulfilling life.

Remember, your energy vibration is a reflection of who you are. Choose to think and feel positive thoughts and you will attract more positivity into your life.

Chapter 10: Aligning your actions with your desires: inspiring inspired action

Introduction

It is not enough to wish for something to get it. We must also take action to achieve our goals. However, many people struggle to take concrete action to achieve their dreams. This is often because their actions are not aligned with their desires.

When our actions are aligned with our desires, we feel motivated and inspired to take action. We have more energy and are more likely to overcome obstacles. When our actions are not aligned with our desires, we feel unmotivated and reluctant to act. We procrastinate and sabotage ourselves often without even realizing it.

Why it is important to align your actions with your desires

Aligning your actions with your desires is important for several reasons:

- It increases your motivation. When you are excited and passionate about something, you are more likely to take action to achieve it.

- It makes you more focused. When you know what you want and why you want it, you are able to focus on your goals and ignore distractions.

- It makes you more resilient. When things get tough, you are more likely to persevere if you are aligned with your deepest desires.

- It leads to better results. When your actions are aligned with your values and goals, you are more likely to achieve success.

How to align your actions with your desires

Here are some tips for aligning your actions with your desires:

- Identify your deepest desires. What do you really want from life? What are your most important values? Take time to reflect on what is really important to you.

- Set clear and concrete goals. Once you know what you want, you need to set clear and concrete goals on how to achieve it. Your goals should be specific, measurable, achievable, relevant and timely (SMART).

- Create an action plan. Once you have established your goals, you need to create an action plan on how to achieve them. Your action plan should break down your goals into smaller, more manageable steps and should include deadlines for each step.

- Take action. The most important thing is to take concrete action toward your goals. Don't wait for the perfect moment; start now. Even small steps can make a difference.

- Be flexible and adapt. Things do not always go as planned, so it is important to be flexible and adapt to changes along the way. Never give up on your goals, but be willing to modify your plan if necessary.

- Celebrate your successes. It is important to celebrate your successes along the way. This will help you stay motivated and focused on your goals.

Exercise: Align your actions with your desires.

Take some time to reflect on your deepest desires. What are the things you really want out of life? What are your most important values?

Once you have a clear understanding of what you want, set a SMART goal and create an action plan on how to achieve it. Take action today and start taking concrete steps toward your dreams.

Aligning your actions with your desires is an ongoing process. It takes time, effort and dedication. But it is worth it. When your actions are aligned with your desires, you are able to live a more meaningful, fulfilling and happy life.

Remember, you are not alone. There are many people who have achieved their goals by aligning their actions with their desires. You can do it too. With the right attitude and the right plan of action, you can achieve whatever you set as your goal.

Chapter 11: Overcoming the fear of failure: embracing learning and growth

Introduction

Fear of failure is one of the most common fears that prevent people from achieving their goals. It can keep us from taking risks, taking on new challenges and pursuing our dreams. If we want to live a full and meaningful life, we must learn to overcome the fear of failure.

Why is the fear of failure so powerful?

Fear of failure is so powerful because it is rooted in our fear of rejection, judgment and failure. We fear being judged by others if we fail and we fear that we are not good enough. This fear can be so debilitating that it prevents us from even trying to achieve our goals.

How fear of failure hinders us

Fear of failure can hinder us in several ways:

- It prevents us from taking risks. When we are afraid of failure, we are less likely to take risks. This can keep us from trying new things and pursuing our dreams.

- It keeps us from taking on new challenges. When we are afraid to fail, we are less likely to challenge ourselves. This can keep us from growing and developing our skills.

- It sabotages us when we are close to success. When we are close to achieving a goal, the fear of failure may take over and sabotage us. We may make mistakes or give up at the last minute.

How to overcome the fear of failure

Here are some suggestions for overcoming the fear of failure:

- Reframe failure. Failure is not the opposite of success; it is an essential part of the success process. Every time we fail, we learn something that we can use to improve the next time.

- Focus on learning, not perfection. Don't try to be perfect. Instead, try to learn and grow from your mistakes.

- Develop a growth mindset. A growth mindset is the belief that you can improve your skills and abilities with effort and dedication.

- Celebrate your small successes. It is important to celebrate your successes along the way, big or small. This will help you stay motivated and focused on your goals.

- Don't be afraid to ask for help. If you are struggling to overcome the fear of failure, don't be afraid to ask a friend, family member, therapist or coach for help.

Exercise: Face your fear of failure.

Think of a time when you were afraid of failure. What held you back? What were the consequences of your fear? What could you have done differently? Now, think of a goal that you are afraid to pursue because of the fear of failure. What are your steps to overcome this fear and start taking action toward your goal?

Overcoming the fear of failure is not easy, but it is possible. When we learn to embrace failure as part of the learning and growth process, we open the door to new possibilities and a fuller, more meaningful life.

Remember, failure is not the end of the world. It is an opportunity to learn and grow. Don't be afraid to fail. Embrace failure and see what you can accomplish.

Chapter 12: Taking advantage of synchronicity and coincidences: seizing the opportunities the universe gives you

Introduction

Synchronicity is a phenomenon that occurs when two or more events occur simultaneously in a meaningful way, even though there is no apparent causal connection between them. Coincidences are events that seem to happen by chance, but may actually have a deeper meaning. Many people believe that synchronicity and coincidences are signs from the universe that we are on the right track or are being guided toward something important.

How synchronicity works

There is no single scientific explanation for synchronicity. However, some theories suggest that it may be due to:

- A universal energy field. Some people believe that there is a universal energy field that connects all things and that synchronicity is a way in which this field communicates to us.

- Our subconscious mind. Other people believe that synchronicity is the way our subconscious mind communicates information to us that we would not otherwise be able to perceive.

- Chance. Some people believe that synchronicity is simply chance, and that there is a rational explanation for every seemingly random event.

How to take advantage of synchronicity and coincidences

If you believe in synchronicity and coincidences, there are some things you can do to take advantage of them:

- Pay attention to signs. The universe may communicate to you through signs such as numbers, symbols, dreams or intuitions. Pay attention to these signs and try to understand what they might mean to you.

- Act on your intuition. If you feel driven to do something, follow your intuition. It could be a sign that the universe is guiding you toward something important.

- Be open to new possibilities. Don't be too attached to your plans. Be open to new possibilities and let the universe guide you.

- Gratitude. Express gratitude for the synchronicities and coincidences that occur in your life. This will help you open yourself to receive even more guidance from the universe.

Exercise: Pay attention to synchronicities.

For the next week, pay attention to the signs of synchronicity in your life. You may notice repeating numbers, symbols you see frequently, or dreams that seem to have special meaning. Take note of these events and try to understand what they might mean for you.

Taking advantage of synchronicity and coincidences can help you live a more meaningful and fulfilling life. When you pay attention to the signs of the universe and act on your intuition, you can open yourself up to new possibilities and realize your dreams.

Remember, the universe is always by your side and is guiding you toward your destiny. Pay attention to the signs, follow your intuition and be open to new possibilities.

Chapter 13: Having faith in the universe: believing that anything is possible

Introduction

Having faith in the universe means believing that everything is possible and that there is a higher force watching over us and guiding us toward our destiny. It means believing that there is a greater plan for our lives, even if we cannot always see it. When we trust the universe, we feel calmer, more confident and optimistic. We are more likely to take risks, pursue our dreams and face challenges with courage.

Why it is important to trust the universe

Having trust in the universe is important for several reasons:

- It reduces stress and anxiety. When we believe that the universe takes care of us, we feel less stressed and anxious. We don't need to worry about everything because we know that things will work out the way they are supposed to.

- It increases our self-esteem. When we believe that the universe has a plan for us, we feel more valuable and important. We know that we have a role to play in the world and that our lives have meaning.

- It makes us more creative and inspired. When we trust the universe, we are more open to new ideas and possibilities. We feel inspired to pursue our dreams and make a difference in the world.

- It helps us overcome difficulties. When things get tough, trusting the universe can help us move forward. We can believe that things will get better and that we can learn and grow from every experience.

How to develop trust in the universe

Here are some suggestions for developing trust in the universe:

- Practice gratitude. Take time each day to express gratitude for the good things in your life. This will help you focus on the positive and attract even more positive things into your life.

- Let go of control. Don't try to control everything. Accept that there are some things that are beyond your control and that it is okay.

- Have faith that things will work out the way they are supposed to. The universe is an intelligent and caring place. Even if things don't always turn out the way we want them to, in the end they happen for our good.

- Listen to your intuition. Your intuition is your inner guide that can help you make better decisions. Pay attention to your intuition and follow it.

- Surround yourself with positive people. The people you spend time with influence the way you think and feel. Surround yourself with positive people who support and encourage you.

Exercise: Strengthen your trust in the universe.

Think of a time when you trusted the universe and things went well. How did that make you feel? What did you learn from that experience? Now, think of a time when you doubted the universe and things didn't turn out as you hoped. How did that make you feel? What did you learn from that experience? The next time you face a difficult challenge, remember the experiences when you trusted the universe and things went well. Use those experiences to strengthen your faith and to believe that things will turn out the way they are supposed to.

Having faith in the universe does not mean that life will always be easy. There will always be challenges and obstacles along the way. But when we trust the universe, we can face them with courage and optimism. We know that we are not alone and that there is a higher force guiding and supporting us.

Remember, the universe is full of love and support. Have faith and let the universe guide you to your destiny.

Chapter 14: Patience is the key: waiting with confidence for your desires to manifest themselves

Introduction

In today's fast-paced world, it is easy to become impatient. We want everything right away, and when things don't go our way, we get frustrated and discouraged. However, patience is an important virtue, especially when it comes to manifesting our desires.

Why patience is important

Patience is important for several reasons:

- It allows us to focus on the present. When we are impatient, we are constantly worried about the future. This prevents us from enjoying the present and living our lives to the fullest.

- It helps us overcome challenges. Things don't always go as planned, and this can be frustrating. But if we are patient, we can face challenges calmly and composedly and find creative solutions.

- It allows us to appreciate the journey. Life is not a destination, but a journey. When we are patient, we can appreciate the journey and learn from our experiences.

- It helps us manifest our desires. When we are impatient, we send signals of negativity and doubt to the universe. This can hinder the manifestation of our desires. But if we are patient, we send signals of faith and trust to the universe, which will help us manifest what we want.

How to develop patience

Here are some suggestions for developing patience:

- Practice gratitude. Take time each day to express gratitude for the good things in your life. This will help you focus on the positive and reduce stress and anxiety.

- Accept that things take time. Don't expect your desires to manifest overnight. Things take time, especially if they are big and important.

- Focus on the process, not the outcome. Don't fixate too much on the end result. Instead, focus on the steps you are taking to achieve your goals.

- Learn from your mistakes. Mistakes are a normal part of life. But instead of viewing them as failures, view them as learning opportunities.

- Be kind to yourself. Don't be too hard on yourself if you feel impatient at times. Everyone does! Just recognize it and get back to practicing patience.

Exercise: Practice patience

Think of a time when you were impatient and it had negative consequences. What did you learn from that experience? Now, think of a time when you practiced patience and it had positive consequences. How did that make you feel? What did you learn from that experience? The next time you feel impatient, remember the experiences when you practiced patience and benefited from it. Use these experiences to remind yourself that patience is an important virtue and worth cultivating.

Patience is not easy, but it is a virtue worth developing. When we are patient, we reduce stress and anxiety, appreciate the present, overcome challenges and manifest our desires. If you are willing to devote time and effort to practice patience, you can improve your life in many ways.

Remember, patience is a sign of strength, not weakness. Don't be afraid to take your time and do things slowly. With patience, you can achieve whatever you set as your goal.

Chapter 15: Gratitude for what you have: attracting even more

Introduction

Gratitude is a powerful emotion that can positively impact our lives in many ways. When we are grateful for the good things in our lives, we feel happier, more optimistic and more fulfilled. We are also more likely to attract even more positive things into our lives.

How gratitude works

Gratitude works by sending positive signals to our brain. These signals help reprogram our minds to focus on the positive, which helps us attract even more positive things into our lives.

How to practice gratitude

There are many ways to practice gratitude. Here are some suggestions:

- Keep a gratitude journal. Every day, take time to write down a few things you are grateful for. They can be big or small things.

- Practice gratitude throughout the day. Take a moment to appreciate the good things in your life, such as a good day, a delicious meal or a pleasant conversation.

- Express gratitude to others. Let people know that you appreciate their presence in your life.

- Focus on the positive side of things. Even when things get tough, try to find the positive side of the situation.

- Practice gratitude even for the little things. The little things in life are often what we take for granted, but they can be a source of great joy and gratitude.

Exercise: Practice gratitude for a week

For the next week, commit to practicing gratitude every day. You can follow the suggestions listed above or find other ways that work for you. At the end of the week, notice how you feel. Have you noticed any change in your mood or perspective?

Gratitude is a simple but powerful way to improve your life. When you practice gratitude, you open yourself up to receiving even more positive things in your life. If you are willing to put in the time and effort to practice gratitude, you may find that your life becomes richer, happier and more fulfilling.

The Invisible Power

Remember, gratitude is not just a feeling, it is a way of life. Choose to be grateful every day and watch how your life is transformed.

Chapter 16: Attracting love and fulfilling relationships

Introduction

We all desire love and fulfilling relationships. However, many people struggle to find them or maintain them. The good news is that there are things we can do to attract love and fulfilling relationships into our lives.

How love and relationship attraction works

Love and relationship attraction works by sending signals to the universe about what we desire. These signals can be sent through our thoughts, emotions and actions. When we send strong and consistent signals about the love and relationships we desire, the universe will help us manifest them.

What to do to attract love and relationships

Here are some suggestions on what to do to attract love and fulfilling relationships into your life:

- Make it clear what you want. Before you can attract love and relationships, you must be clear about what you want. What qualities do you want in a partner? What kind of relationship do you desire?

- Visualize what you desire. Take time each day to visualize yourself in a happy and fulfilling relationship. Imagine what it would be like, how you would feel, and what you would do together.

- Affirm your beliefs. Create positive affirmations about love and relationships. Repeat these affirmations every day with conviction and feeling.

- Be open and available. You cannot attract love and relationships if you are closed and unavailable. Open your heart to love and be willing to put yourself out there.

- Love yourself. The most important person to love is you. When you love yourself, you radiate love and naturally attract people who love and respect you.

- Let go of the past. If you are still clinging to past relationships, you will not be able to open up to new relationships. Let go of the past and forgive yourself and others.

- Trust the universe. The universe wants you to be happy and have fulfilling relationships. Have faith that the universe will guide you to the right person for you.

Exercise: Create a visionary bulletin board for love

Create a visionary bulletin board for love. Collect pictures, words and phrases that represent love and the relationships you desire. Hang your visionary bulletin board in a place where you can see it often. This will help you stay focused on your goals and attract love and positive relationships into your life.

Attracting love and fulfilling relationships takes time, effort and patience. But if you are willing to dedicate yourself to this process, you can create the love life you have always wanted. Remember, you deserve to be loved and happy.

Some additional suggestions:

- Be yourself. Don't try to be someone you are not in order to attract a partner. The right person will love you for who you are.

- Be honest. Honesty is essential in any relationship. Be honest with yourself and your partner about your wants and needs.

- Communicate openly. Communication is the key to any healthy relationship. Talk openly and honestly with your partner about your thoughts, feelings and needs.

- Respect your partner. Every person deserves to be treated with respect. Respect your partner even when you disagree with him or her.

- Be considerate. Do nice things for your partner to show him or her how much you care about him or her.

- Forgive. Everyone makes mistakes. Learn to forgive your partner when they make mistakes.

- Be willing to compromise. There is no such thing as a perfect couple. Both partners must be willing to compromise to make a relationship work.

- Have fun together! Life is too short to be serious all the time. Make sure you have fun with your partner.

If you follow these tips, you will be well on your way to building loving and fulfilling relationships.

Chapter 17: Manifesting financial abundance and prosperity

Introduction

We all desire to have financial abundance and live a prosperous life. But how can we manifest it? The good news is that there are things we can do to increase our chances of financial success.

How the manifestation of financial abundance works

The manifestation of financial abundance works by sending signals to the universe about what we desire. These signals can be sent through our thoughts, our emotions and our actions. When we send strong and consistent signals about the financial abundance we desire, the universe will help us manifest it.

What to do to manifest financial abundance

Here are some suggestions on what to do to manifest financial abundance and prosperity in your life:

The Invisible Power

1. Set a clear financial goal.

What does "financial abundance" mean to you? How much money do you want to earn? What lifestyle do you want to live? Once you have a clear goal in mind, you can start focusing on how to achieve it.

2. Develop an abundance mindset.

An abundance mindset is the belief that there are infinite resources available to you and that you deserve to be rich and prosperous. If you have a scarcity mentality, you will believe that money is limited and that you will have to struggle to get it. This negative attitude will hinder your efforts to manifest financial abundance.

3. Visualize your success.

Take time each day to visualize yourself in a situation of financial abundance. Imagine what it would be like to have all the money you desire and how you would feel. The more vivid and detailed your visualization, the better.

4. Affirm your beliefs.

Create positive affirmations about financial abundance. Repeat these affirmations every day with conviction and feeling. For example, you might affirm, "I am a magnet for money. Money flows to me easily and effortlessly. I am grateful for my financial abundance."

5. Take action to achieve your goals.

It is not enough just to dream and visualize. You must also take action to achieve your financial goals. Create a plan and take concrete action to achieve it. This could include finding a better job, starting a business, or investing in stocks or real estate.

6. Be grateful for what you have.

Gratitude is a powerful force that can attract even more abundance into your life. Take time each day to express gratitude for the money you have, no matter how little.

7. Release limiting beliefs.

You may have limiting beliefs about money that prevent you from manifesting financial abundance. For example, you may believe that "money does not make you happy" or that "I am not good with money." Identify your limiting beliefs and challenge them. Replace them with positive, reinforcing beliefs.

8. Give to others.

One of the best ways to attract financial abundance is to give to others. Donate to charity or help people in need. When you give, you are sending a signal to the universe that you are open to receiving.

9. Be patient.

Manifesting financial abundance takes time and patience. Don't get discouraged if you don't see immediate results. Keep focusing on your goals and taking action, and eventually you will achieve success.

10. Have fun!

The more fun you have pursuing your financial goals, the more likely you are to achieve them. Money and prosperity should be a source of joy, not stress.

Exercise: Create a journal of abundance

Create an abundance journal and use it to write down your financial goals, positive affirmations and experiences of gratitude. Also write down your thoughts and feelings about money and prosperity. This will help you stay focused on your goals and monitor your progress.

Manifesting financial abundance is possible, but it requires commitment and perseverance. If you follow these tips, you will be well on your way to creating the prosperous life you desire.

Remember, you deserve to be rich and happy. Have faith in your abilities and never give up on your dreams.

Chapter 18: Achieving your career goals and success

Introduction

We all desire to have a rewarding and successful career. But how can we achieve it? The good news is that there are things we can do to increase our chances of professional success.

How career goal setting works

Achieving career goals works by sending signals to the universe about what we desire. These signals can be sent through our thoughts, our emotions, and our actions. When we send strong and consistent signals about the career goals we desire, the universe will help us manifest them.

What to do to achieve your career goals

Here are some suggestions on what to do to achieve your career goals and success:

1. Identify your goals.

What do you want to achieve in your career? What is your ideal job? What are your long-term goals? Once you have a clear idea of what you want, you can begin to create a plan to achieve it.

2. Develop your skills and knowledge.

What skills and knowledge do you need to achieve your career goals? Make sure you have the skills and knowledge you need to succeed in your field. This could include obtaining an education, attending training courses, or acquiring certifications.

3. Create a network of contacts.

Relationships are important in any career. Create a network of contacts with people in your field. Attend networking events, join a professional association, or directly contact people you admire.

4. Be proactive.

Don't wait for things to happen to you. Take control of your career and take the initiative. Look for new opportunities, apply for jobs that interest you, and show off in the workplace.

5. Be resilient.

The path to success is not always easy. There will be obstacles and setbacks along the way. But it is important to be resilient and never give up on your dreams. Learn from your mistakes, get back up and keep moving forward.

6. Have a positive outlook.

Your attitude is important to your success. Keep a positive outlook and focus on your goals. Believe in yourself and your abilities.

7. Be willing to take risks.

If you are not willing to take risks, you will never achieve great things. Get out of your comfort zone and be willing to put yourself on the line.

8. Be patient.

Success does not happen overnight. It takes time, commitment and perseverance to achieve your goals. Be patient and never give up.

9. Have fun.

Work should be rewarding, not stressful. Find work that you are passionate about and that you enjoy. When you have fun, you are more likely to succeed.

10. Give to others.

One of the best ways to be successful is to help others. Volunteer in your field, mentor someone who is starting their career or share your knowledge with others.

Exercise: Create a career plan

Create a career plan that outlines your short- and long-term goals. Your plan should include the steps you need to take to achieve your goals, as well as the deadlines and resources needed. Review your career plan regularly and make changes if necessary.

Achieving your career goals and success is possible, but it requires commitment and dedication. If you follow these tips, you will be well on your way to achieving your dreams.

Remember, you have the potential to achieve great things. Believe in yourself and never give up on your goals.

Some additional suggestions:

- Find a mentor. A mentor can offer you valuable guidance, support and advice.

- Attend training and professional development courses. This will help you stay current on the latest trends and develop your skills.

- Read books and articles about your field. This will help you learn about your field and stay inspired.

- Attend industry events. This will help you meet people in your field and stay up-to-date on the latest news.

- Be a team player. Be a team player

Chapter 19: Improving your health and well-being

Introduction

Health and well-being are essential to living a happy and fulfilling life. When we feel good physically and mentally, we have more energy, are more productive, and are able to enjoy life more.

How to improve your health and well-being

There are many things you can do to improve your health and well-being. Here are some suggestions:

1. Eat a healthy diet.

What you eat has a direct impact on your health and well-being. Choose fresh, whole foods such as fruits, vegetables, whole grains, and lean proteins. Limit your intake of processed foods, foods high in sugar and saturated fats.

2. Exercise regularly.

Exercise is important for physical and mental health. Try to get at least 30 minutes of moderate exercise most days of the week. You can walk, run, swim, bike or do any other activity you enjoy.

3. Get enough sleep.

Sleep is essential for the body and mind to rest and recharge. Most adults need about 7 to 8 hours of sleep per night.

4. Manage stress.

Stress can have a negative impact on your physical and mental health. Find healthy ways to manage stress, such as yoga, meditation or spending time in nature.

5. Avoid smoking and alcohol abuse.

Smoking and alcohol abuse can seriously damage your health. If you smoke or abuse alcohol, talk to your doctor about getting help to quit.

6. Get regular medical checkups.

Regular medical checkups can help you identify and treat any health problems early. Be sure to follow your doctor's recommendations for routine checkups.

7. Cultivate positive relationships.

Positive relationships with family, friends, and loved ones are important for your mental health and well-being. Spend time with people who make you feel good and supportive.

8. Do something you enjoy.

Make time to do things that you enjoy and that make you happy. This could be anything from reading to gardening to sports or spending time with your loved ones.

9. Help others.

Helping others is a great way to feel good about yourself and to make a difference in the world. Volunteer for a cause that interests you or find other ways to help people in need.

10. Be grateful for the good things in your life.

Gratitude can help you focus on the positive aspects of your life and feel happier and more fulfilled. Every day, take time to express gratitude for the good things in your life.

Exercise: Create a plan for healthy living

Create a plan for healthy living that includes healthy foods, regular exercise, sufficient sleep and ways to manage stress. Your plan should also include activities you enjoy and ways to connect with others. Review your plan regularly and make changes if necessary.

Improving your health and well-being is an ongoing journey. But by making small positive changes in your life, you can make a big difference in your overall health and well-being.

Remember, you deserve to feel good. Take care of yourself and make the best choices for your health and well-being.

Some additional suggestions:

- Listen to your body. If you feel tired, sore or stressed, slow down and take time for yourself.

- Choose joy. Do things that make you happy and feel good.

- Be kind to yourself. Everyone makes mistakes. Learn from your mistakes and move on.

- Live in the present moment. Don't dwell on the past or worry about the future. Focus on living in the present moment.

- Appreciate the beauty around you. Take time to appreciate the beauty of nature, art and music.

- Practice compassion. Be kind and understanding to yourself and others.

**Make a difference in the world.

Chapter 20: Creating a life rich in positive experiences

Introduction

We all desire a life rich in positive experiences. But how can we create it? The good news is that there are things we can do to increase our chances of living a life full of joy, love and happiness.

How to create a life full of positive experiences

Here are some tips on how to create a life rich in positive experiences:

1. Focus on the positive.

What we focus on in our lives tends to grow. If we focus on negative things, we will feel negative. But if we focus on positive things, we will feel more positive and attract more positive experiences into our lives.

2. Be grateful for the good things in your life.

Gratitude is a powerful force that can help us focus on the positive and attract even more positive things into our lives. Every day, take time to express gratitude for the good things in your life.

3. Spend time with positive people.

The people we spend time with have an impact on us. If we spend time with positive and supportive people, we will feel more positive and happy. But if we spend time with negative and critical people, we will feel more negative and stressed.

4. Do things you enjoy.

Make time to do things you enjoy and that make you feel good. This could be anything from reading to gardening to sports or spending time with your loved ones.

5. Get out of your comfort zone.

Trying new things and stepping out of your comfort zone can help you grow and have new experiences. Don't be afraid to take risks and try new things.

6. Travel and explore the world.

Traveling and exploring the world is a great way to learn about new cultures, experience new things, and create lasting memories.

7. Help others.

Helping others is a great way to feel good about yourself and to make a difference in the world. Volunteer for a cause that interests you or find other ways to help people in need.

8. Be open to new experiences.

Life is full of opportunities for new experiences. Be open to new possibilities and don't be afraid to try new things.

9. Live in the present moment.

Don't dwell on the past or worry about the future. Focus on living in the present moment and appreciate every moment.

10. Follow your intuition.

Your intuition is a powerful guide that can help you make positive decisions and live a fulfilling life. Listen to your intuition and follow your heart.

Exercise: Create a wish list

Create a wish list of all the things you would like to do and experience in your life. Your wish list can include anything from traveling to exotic destinations to learning new skills to meeting interesting people. Don't limit yourself! The bigger and more ambitious your wish list is, the better.

Creating a life full of positive experiences is an ongoing journey. But by taking small steps each day, you can begin to create the life you've always wanted.

Remember, life is short and precious. Don't waste it by living a life of regret. Live your life to the fullest and create experiences you will remember forever.

Some additional suggestions:

- Be present in the moment. When you are with someone, give them your full attention. Don't think about anything else or check your phone.

- Appreciate the little things. The little things in life are often the most important. Take time to appreciate simple things like a beautiful sunset, a hot cup of coffee, or a laugh with a friend.

- Forgive yourself and others. Everyone makes mistakes. Learn from your mistakes and move on. And forgive others when they make mistakes.

- Be kind to yourself and to others. Kindness is one of the most important qualities you can possess. Be kind to yourself and to others, even when things are difficult.

- Accept what you cannot change. Some things in life are out of our control. Learn to accept what you cannot change and focus on what you can control.

- Focus on your own path.

Chapter 21: Making a positive impact on the world

Introduction

We all wish to have a positive impact on the world. But how can we do this? The good news is that there are many things we can do, big or small, to make a difference in the lives of others and make the world a better place.

How to make a positive impact on the world

Here are some suggestions on how to make a positive impact on the world:

1. Be kind to others.

Kindness is one of the simplest things you can do to make a positive impact on the world. Smile at people, hold the door open for someone, or offer your help to those in need. These small acts of kindness can make a big difference in someone's day.

2. Volunteer for a cause that matters to you.

There are many organizations that need volunteers to help them accomplish their goals. Find a cause that interests you and donate your time to make a difference.

3. Donate to charity.

Even a small donation can make a big difference to a charitable organization that is working to solve important problems. If you have the means, consider donating to charity to a cause that matters to you.

4. Spread awareness about important issues.

If there is an issue you care about, speak up and make your voice heard. You can write letters to your elected representatives, participate in peaceful protests, or spread awareness on social media.

5. Respect the environment.

There are many things you can do to reduce your impact on the environment. Recycle and reuse, conserve water and energy, and choose environmentally friendly products.

6. Be a good example.

The best way to inspire others to make a difference is to be a good example to them. Live your life in a way that is consistent with your values and show others how they can make a positive impact on the world.

7. Be open and tolerant.

The world is a place full of diversity. Be open to learning about different cultures and perspectives. Be tolerant of people who are different from you.

8. Promote peace and understanding.

Peace and understanding are essential for a better world. Work to promote peace in your community and in the world.

9. Be creative.

There are many creative ways to make a positive impact on the world. Use your creativity to find new ways to make a difference.

10. Never give up.

Changing the world is not easy. There will be setbacks and obstacles along the way. But never give up on your dreams. Keep working to make the world a better place.

The Invisible Power

Exercise: Create a plan to change the world

Think about a problem that matters to you and how you could help solve it. Create a plan that describes what you will do to make a positive impact. Your plan can include anything from volunteering for a local organization to launching your own initiative.

Making a positive impact on the world is possible for everyone. Even the smallest gestures can make a difference. If everyone did their part, we could make the world a much better place.

Remember, you have the power to make a difference. Start today to create a positive impact on the world.

Some additional suggestions:

- Find a group of people who share your values and work together to make a difference.

- Be patient. Change takes time.

- Don't be afraid to fail. Failure is part of the learning process.

- Celebrate your successes.

- Have fun! Making a difference should be rewarding.

Chapter 22: Staying motivated when things get tough

Introduction

We all go through difficult periods in our lives. There are times when we feel unmotivated, discouraged, and as if we are not making progress. But it is important to remember that we are not alone. Everyone goes through these challenges, and the important thing is to find ways to stay motivated and keep moving forward.

How to stay motivated when things get tough

Here are some tips on how to stay motivated when things get tough:

1. Remember your goal.

Why are you doing this? What is your goal? Having a clear and defined goal can help you stay focused and motivated, even when things get tough.

2. Break your goals down into smaller goals.

A large goal may seem insurmountable, which can lead to demotivation. Break your goals down into smaller, more manageable goals. This will make the process less daunting and help you stay motivated along the way.

3. Celebrate your successes.

No matter how small they may seem, take the time to celebrate your successes. This will help you stay focused on the progress you are making and motivate you to continue.

4. Find a mentor or role model.

Having someone to look up to and learn from can be a great source of motivation. Find a mentor or role model who has achieved what you want to achieve and learn from their experience.

5. Surround yourself with positive people.

The people we spend time with have an impact on us. Surround yourself with positive and supportive people who believe in you and your dreams.

6. Don't be afraid to fail.

Failure is part of the learning process. Don't be afraid to fail, but learn from your mistakes and move on.

The Invisible Power

7. Focus on the present.

Don't dwell on the past or worry about the future. Focus on the present and do your best in the present moment.

8. Take care of yourself.

Make sure you get enough sleep, eat healthy foods and exercise regularly. Taking care of yourself physically and emotionally will help you stay motivated and focused.

9. Do something you enjoy.

Make time to do things you enjoy and that make you happy. This will help you reduce stress and stay positive.

10. Never give up.

The most important thing is to never give up on your dreams. Keep working hard and you will eventually reach your goals.

Exercise: Create a vision board.

A vision board is a collage of pictures and words that represent your goals and dreams. Creating a vision board can help you stay focused and motivated. You can hang your vision board in a place where you see it often, such as your desk or bedside table.

Staying motivated when things get tough is a challenge, but it is possible. By following these tips, you can stay focused on your goals and keep moving forward, even when things get tough.

Remember, you have the power to achieve your dreams. Never give up!

Some additional suggestions:

- Listen to motivational music.

- Read inspirational books and articles.

- Watch motivational videos.

- Attend motivational seminars and conferences.

- Find a support group online or in person.

- Talk to a therapist or counselor.

Remember, you are not alone. There are many people who have gone through what you are going through and come out successful. With hard work, dedication, and the right motivation, you too can achieve your goals.

Chapter 23: Dealing with internal doubts and resistance

Introduction

We all face internal doubts and resistances throughout our lives. These mental barriers can hinder our progress and prevent us from achieving our goals. However, it is important to remember that we are not alone in this struggle. There are effective strategies for dealing with these obstacles and unlocking our full potential.

How to deal with internal doubts and resistance

Here are some suggestions on how to deal with internal doubts and resistance:

1. Identify your fears and doubts.

The first step in overcoming internal doubts and resistances is to identify their root. What is holding you back? What are your biggest fears? Once you know what you are facing, you can begin to develop strategies to overcome it.

2. Question your negative thoughts.

Our thoughts have a powerful impact on our behavior and emotions. When we focus on negative thoughts, we are more likely to feel unmotivated and discouraged. Learn to question your negative thoughts and replace them with more positive and realistic thoughts.

3. Challenge your limiting beliefs.

We often limit ourselves because of limiting beliefs we have about ourselves or the world around us. Identify your limiting beliefs and challenge them with concrete evidence. What makes you believe that you cannot achieve your goals? What evidence suggests otherwise?

4. Visualize your success.

Visualization is a powerful technique that can help you overcome internal doubts and resistance. Take time each day to visualize yourself achieving your goals. Imagine how you will feel when you have achieved success. This will help you stay focused and motivated.

5. Surround yourself with positive people.

The people we spend time with have an impact on us. Surround yourself with positive and supportive people who believe in you and your dreams. Avoid negative people who make you doubt yourself.

6. Celebrate your successes.

No matter how small they may seem, take time to celebrate your successes. This will help you stay focused on the progress you are making and motivate you to continue.

7. Don't be afraid to fail.

Failure is part of the learning process. Don't be afraid to fail, but learn from your mistakes and move forward. Every failure is an opportunity to grow and improve.

8. Be patient.

Overcoming internal doubts and resistance takes time and patience. Don't expect to change overnight. Keep working on yourself and you will eventually reach your goals.

9. Seek professional help.

If you are struggling to overcome your doubts and internal resistance on your own, don't hesitate to seek professional help. A therapist or counselor can provide you with the tools and support you need to overcome these obstacles.

Exercise: Write a journal about your thoughts and feelings

Keeping a journal is a great way to explore your thoughts and feelings. Writing regularly about your internal doubts, fears and resistances can help you identify and understand them better. It can also be helpful to write about your goals and dreams and how you plan to achieve them.

Dealing with internal doubts and resistances is a challenge, but it is possible. By following these tips, you can develop the mental strength and resilience you need to overcome these obstacles and reach your full potential.

Remember, you have the power to create the life you desire. Don't let internal doubts and resistance hold you back.

Some additional suggestions:

- Practice mindfulness and meditation.

- Do physical activity regularly.

- Eat healthy and nutritious foods.

- Get enough sleep.

- Learn how to manage stress.

- Focus on the present.

- Be grateful for the good things in your life.

- Never give up!

**Overcoming internal doubts and resistance is a journey of growth and transformation. With commitment and perseverance, you can unlock your true potential and live by getting the most out of yourself.

Chapter 24: Dealing with setbacks and obstacles

Introduction

Life is full of ups and downs. There will be times when everything goes according to plan and others when we face setbacks and obstacles. These unexpected events can be frustrating and discouraging, but it is important to remember that they are not the end of the world. With the right approach, we can overcome these challenges and come out stronger and more resilient.

How to handle setbacks and obstacles

Here are some tips on how to handle setbacks and obstacles:

1. Accept the situation.

The first thing to do when you face a setback or obstacle is to accept it. Denying the reality or fighting against it will only make things worse. Accept the situation for what it is and focus on solving it.

2. Keep calm.

It is easy to panic when things do not go as planned. However, it is important to remain calm and focused. Take a few deep breaths and remember that you are capable of dealing with this challenge.

3. Assess the situation.

Once you have calmed down, take some time to assess the situation. What exactly happened? What are the causes of the problem? What are your options? Once you have a better understanding of the situation, you can begin to develop a plan to resolve it.

4. Look for creative solutions.

Don't limit yourself to the obvious solutions. Often, the best ways to overcome obstacles are the most creative ones. Think outside the box and look for solutions that may not have occurred to you at first.

5. Ask for help.

Don't be afraid to ask others for help. They may have advice or resources that can help you overcome the challenge. Talk to friends, family, colleagues or a mentor. You may also want to consider seeking professional help from a therapist or counselor.

6. Focus on the present.

It is easy to dwell on what went wrong or worry about what the future holds. However, it is important to focus on the present and what you can control. Do your best to deal with the challenge in front of you and let the past and future take care of themselves.

7. Learn from your mistakes.

Every setback and obstacle is an opportunity to learn and grow. Reflect on what went wrong and how you could have done things differently. Use this knowledge to improve your ability to cope with future challenges.

8. Be patient.

Overcoming setbacks and obstacles takes time and patience. Don't expect to solve everything overnight. Do your best every day and you will eventually reach your goals.

9. Keep a positive outlook.

Even when things get tough, it is important to keep a positive outlook. Remember that you have the power to overcome this challenge and that you will come out stronger on the other side.

10. Celebrate your successes.

No matter how small they may seem, take time to celebrate your successes. This will help you stay motivated and focused on achieving your goals.

Exercise: Create a plan for dealing with setbacks.

Think of a setback or obstacle you have faced in the past. Create a plan on how you could have handled the situation differently. Your plan should include the following steps:

- Assess the situation

- Look for creative solutions

- Ask for help

- Focus on the present

- Learn from mistakes

- Being patient

- Maintaining a positive outlook

- Celebrating successes

Dealing with setbacks and obstacles is an inevitable part of life. However, with the right approach, we can overcome these challenges and come out stronger and more resilient.

Remember, you have the power to overcome any obstacle. Never give up!

Some additional suggestions:

- Practice gratitude.

- Take care of yourself physically and emotionally.

- Do something you enjoy.

- Surround yourself with positive people.

- Never give up!

**Overcoming setbacks and obstacles is a journey of growth and transformation. With commitment and perseverance, you can unlock to positivity and attract everything you want.

Chapter 25: Learning from failures and rising up stronger

Introduction

Failure is an inevitable part of life. Everyone makes mistakes and goes through difficult times. However, how we deal with failure determines our future success. If we learn from our mistakes and get back up stronger, failure can become a stepping stone to success.

How to learn from failure and get back up stronger

Here are some tips on how to learn from failures and stand up stronger:

1. Accept failure.

The first thing to do when you fail is to accept it. Denying reality or fighting against it will only make things worse. Accept failure for what it is and focus on learning from it.

2. Don't take it personally.

It is easy to feel discouraged and unmotivated when you fail. However, it is important to remember that failure does not mean you are a failure. Don't take it personally and focus on what you can do to improve in the future.

3. Identify the cause of failure.

Once you have accepted failure, it is important to identify its cause. What went wrong? What were your mistakes? Once you know what went wrong, you can begin to develop a plan to avoid making the same mistakes in the future.

4. Learn from your mistakes.

Failure is an opportunity to learn and grow. Reflect on what went wrong and how you could have done things differently. Use this knowledge to improve your ability to cope with future challenges.

5. Develop a plan for the future.

Once you have learned from your mistakes, it is important to develop a plan for the future. What do you want to do differently next time? What are your goals? Having a plan will help you stay focused and motivated.

6. Don't be afraid to try again.

Failure does not mean you have to give up on your dreams. Don't be afraid to try again and again until you reach your goals.

7. Seek inspiration from others.

Many great successes have failed many times before achieving success. Seek inspiration from the stories of other people who have overcome failure and achieved their goals.

8. Keep a positive outlook.

Even when things get tough, it is important to keep a positive outlook. Remember that you have the power to overcome any obstacle.

9. Celebrate your successes.

No matter how small they may seem, take time to celebrate your successes. This will help you stay motivated and focused on achieving your goals.

10. Never give up!

The most important thing is to never give up on your dreams. Keep working hard and you will eventually reach your goals.

Exercise: Write a journal about your failures

Keeping a journal is a great way to reflect on your failures and learn from them. Write about your past failures and what you learned from them. This will help you develop a more resilient mindset and set yourself up for future success.

Learning from failures and getting back up stronger is a process that takes time and commitment. However, it is worth it. If you learn from your mistakes and never give up on your dreams, you can achieve anything you set your mind to.

Remember, you have the power to create the life you desire. Don't let failure hold you back.

Some additional suggestions:

- Practice gratitude.

- Take care of yourself physically and emotionally.

- Do something you enjoy.

- Surround yourself with positive people.

- Never give up!

Overcoming failures and challenges is a journey of growth and transformation. With commitment and perseverance, you can unlock your true potential and live the life you desire.

Chapter 26: Celebrating your successes and keeping your focus

Introduction

Achieving your goals requires hard work, dedication and perseverance. But it is equally important to celebrate your successes along the way. Taking the time to recognize your accomplishments can help you stay motivated, focused, and inspired to keep growing and improving.

How to celebrate your successes and stay focused

Here are some tips on how to celebrate your successes and keep focused:

1. Acknowledge your successes.

The first thing to do is to recognize your successes. No matter how small they may seem, take the time to appreciate what you have accomplished. This will help you feel good about yourself and motivate you to continue to achieve your goals.

2. Celebrate your accomplishments in a way that is meaningful to you.

There are many different ways to celebrate your successes. Find a way that is meaningful to you and makes you feel good. You could have a party with friends and family, treat yourself to a special gift, or simply take time to relax and enjoy your success.

3. Reflect on what you have learned along the way.

Achieving your goals is not just about the end result. It is also about what you learn along the way. Reflect on what you have learned from your experiences and how you can use this knowledge to improve in the future.

4. Set new goals.

Once you have celebrated your success, it is time to set new goals. This will help you stay focused and motivated to keep growing and improving.

5. Visualize your success.

Visualization is a powerful technique that can help you stay focused on your goals. Take time each day to visualize yourself achieving your goals. Imagine how you will feel when you have achieved success. This will help you stay motivated and focused.

6. Surround yourself with positive people.

The people we spend time with have an impact on us. Surround yourself with positive and supportive people who believe in you and your dreams. Avoid negative people who make you doubt yourself.

7. Practice gratitude.

Gratitude is a powerful tool that can help you stay focused on the positive in your life. Take time each day to express gratitude for the good things in your life. This will help you stay positive and motivated.

8. Take care of yourself.

Make sure you get enough sleep, eat healthy foods and exercise regularly. Taking care of yourself physically and emotionally will help you stay focused and motivated to achieve your goals.

9. Don't be afraid to fail.

Failure is part of the learning process. Don't be afraid to fail, but learn from your mistakes and move on. Every failure is an opportunity to grow and improve.

10. Never give up!

The most important thing is to never give up on your dreams. Keep working hard and you will eventually reach your goals.

Exercise: Create a success board

A success board is a great way to keep track of your successes and stay motivated. Create a success board and add photos, newspaper clippings, or other mementos that represent your accomplishments. Look at your success board every day to remind yourself how far you have come and to inspire you to keep growing and improving.

Celebrating your successes and staying focused is an important part of achieving your goals. By taking the time to recognize your accomplishments, you will stay motivated, focused and inspired to keep growing and improving.

Remember, you have the power to create the life you desire. Don't let anything hold you back.

Some additional suggestions:

- Reward yourself for your successes.

- Share your successes with others.

- Join a support group or coaching program.

- Find a mentor or role model.

- Never give up!

**Overcoming challenges and achieving your goals is a journey of growth and transformation.

Chapter 27: The power of attraction: a journey of continuous transformation

Introduction

The power of attraction is a fascinating concept that has gained increasing popularity in recent years. It is based on the idea that our thoughts, emotions and beliefs can influence our reality. In other words, what we give attention to and focus on tends to attract similar experiences and opportunities to itself.

How to harness the power of attraction for continued transformation

Here are some suggestions on how to harness the power of attraction for continuous transformation:

1. Clarify your goals and desires.

The first step to attracting what you desire into your life is to clarify your goals and desires. What do you really want to achieve? What do you aspire to? Take some time to reflect on what is really important to you and write down your goals clearly and concisely.

2. Visualize your successes.

Visualization is a powerful technique that can help you program your subconscious mind for success. Take time each day to visualize yourself achieving your goals. Imagine how you will feel when you have achieved success. This will help you stay motivated and focused.

3. Affirm your positive beliefs.

Our beliefs have immense power over our lives. If we believe we can achieve something, we are more likely to do so. Conversely, if we believe we cannot achieve something, we are less likely to even try. This is why it is important to regularly affirm positive beliefs about yourself and your goals.

4. Be grateful for what you have.

Gratitude is a powerful emotion that can help you focus on the positive in your life. Take time each day to express gratitude for the good things in your life. This will help you attract even more positive things into your life.

5. Make room for new experiences and opportunities.

To attract new experiences and opportunities into your life, it is important to make room for them. This means getting rid of negative thoughts, emotions and beliefs that may be holding you back. It also means being open to new experiences and opportunities, even if they do not fit perfectly into your plans.

6. Act with faith and determination.

The power of attraction does not work by itself. It is important to act with faith and determination to achieve your goals. Don't expect things to simply happen to you. Go out and get what you want out of life.

7. Be patient and persistent.

The power of attraction does not work overnight. It takes time, patience and perseverance to see tangible results. Don't get discouraged if you don't see immediate results. Keep focusing on your goals and taking action with faith, and eventually you will achieve them.

8. Learn from your mistakes and failures.

Failure is an inevitable part of life. But failure is not the end of the world. It is an opportunity to learn and grow. Learn from your mistakes and failures and use them as a springboard to success.

9. Never give up!

The most important thing is to never give up on your dreams. Keep working hard and you will eventually reach your goals.

Exercise: Create a gratitude journal

A gratitude journal is a great way to focus on the positive in your life and attract even more positive things. Get a notebook and write down three things each day that you are grateful for. It can be anything, big or small. The more you focus on the good things in your life, the more you will attract.

The power of attraction is a powerful tool that can help you create the life you desire. However, it is important to remember that it is not a magic wand. It requires commitment, patience and perseverance. If you are willing to put in the work, the power of attraction can help you achieve your biggest dreams.

Remember, you have the power to create the life you desire. Don't let anything hold you back.

Some additional suggestions:

- Surround yourself with positive people.

- Practice mindfulness and meditation.

- Take care of yourself physically and emotionally.

- Do something you enjoy.

**Never give up.

Chapter 28: Sharing the power of attraction with others

Introduction

The power of attraction is not only a concept to be applied in one's own life, but it can also be shared with others to create a positive impact in the world. By helping others understand and use the power of attraction, we can help raise the collective consciousness and create a more positive reality for all.

How to share the power of attraction with others

Here are some suggestions on how to share the power of attraction with others:

1. Talk about the power of attraction with people you know.

Share your experiences with the power of attraction with friends, family and colleagues. Tell them how you used the power of attraction to achieve your goals and how it improved your life.

2. Organize workshops or seminars on the power of attraction.

If you have a thorough understanding of the power of attraction, you can organize workshops or seminars to teach others how to use it in their lives. This is a great way to spread awareness and help people create positive changes in their lives.

3. Create online content about the power of attraction.

You can create a blog, YouTube channel or social media account to share information about the power of attraction with a wider audience. Write articles, create videos or record podcasts that teach the key concepts of the power of attraction and how to apply them in daily life.

4. Offer coaching or mentoring on the power of attraction.

If you have experience in helping people achieve their goals using the power of attraction, you can offer coaching or mentoring to those who want to deepen their understanding and application of this principle.

5. Support organizations that promote the power of attraction.

There are many organizations working to spread awareness about the power of attraction and to help people use it to create a better life. You can support these organizations by donating, volunteering or spreading their word.

Remember, the power of attraction is a powerful tool that can be used to create positive change in the world. By sharing it with others, you can help raise the collective consciousness and create a more loving, compassionate and abundant reality for all.

In addition to the tips mentioned above, here are some other tips to keep in mind when sharing the power of attraction with others:

- Be authentic and passionate. People will be more likely to listen to you if they see that you are truly passionate about the power of attraction and believe in its potential to create positive change.

- Focus on the positive. When you talk about the power of attraction, focus on the positive aspects and the potential rewards it can bring to people's lives. Avoid dwelling on the negative aspects or potential failures.

- Be respectful of other people's beliefs. Do not try to force your beliefs on others. Everyone has their own learning path and should be free to explore the power of attraction at their own pace.

- Be patient. The power of attraction is not a quick fix. It takes time and commitment to see tangible results. Be patient with yourself and others as you learn and apply this principle.

- Have fun! The power of attraction should be an enjoyable and stimulating journey. Have fun exploring this powerful concept and sharing your discoveries with others.

Sharing the power of attraction with others is a valuable way to help create a better world. By using the above tips and advice, you can make a difference in people's lives and help them create the life they desire.

I would love to hear what you think of my book. Would you be able to leave a review?

Thank you, Joseph La Belle.

Grab Your Free Bonus eBook Now!

Part 2: Acting and Receiving

- Chapter 10: Aligning Your Actions with Your Desires: Inspiring Action
- Chapter 11: Overcoming Fear of Failure: Embracing Learning and Growth
- Chapter 12: Exploiting synchronicity and coincidences: seizing the opportunities the universe offers you
- Chapter 13: Trusting the Universe: believing that anything is possible
- Chapter 14: Patience is the Key: Waiting with confidence for your desires to manifest
- Chapter 15: Gratitude for What You Have: Attracting Even More

Part 3: Applying the Power of Attraction in Different Areas of Life

- Chapter 16: Attracting Love and Fulfilling Relationships
- Chapter 17: Manifesting Financial Abundance and Prosperity
- Chapter 18: Achieve Your Career and Success Goals
- Chapter 19: Improving Your Health and Well-Being
- Chapter 20: Creating a Life Rich in Positive Experiences

- Chapter 21: Making a Positive Impact on the World

Part 4: Staying Motivated and Overcoming Challenges

- Chapter 22: Staying Motivated When Things Get Tough
- Chapter 23: Dealing with Internal Doubts and Resistance
- Chapter 24: Dealing with setbacks and obstacles
- Chapter 25: Learn from failures and stand up stronger
- Chapter 26: Celebrating your successes and maintaining focus

Conclusion

- Chapter 27: The Power of Attraction: A Journey of Continuous Transformation
- Chapter 28: Sharing the Power of Attraction with Others

Chapter 1: The power of attraction: what it is and how it works

Introduction

The power of attraction is a universal law that states that **like attracts like**. In other words, our thoughts, emotions and beliefs create an energy field that attracts experiences and people in harmony with that vibration to us.

This principle has been expressed in different cultures and traditions for centuries, often under different names such as the 'law of magnetism' or the 'law of the mind'. However, it is only in recent decades that the power of attraction has begun to receive scientific attention and be applied in practical ways to improve people's lives.

How does the power of attraction work?

In principle, the power of attraction works through three simple steps:

1. **Ask:** Clearly define what you want to attract into your life. This can be done through setting goals, creating vision boards or simply

focusing on what you want with clarity and intensity.

2.   **Belief:** Have an unshakeable belief that it is possible to achieve what you want. This belief does not have to be based on concrete evidence, but rather on an inner trust that the universe will provide us with what we need.

3.   **Receive:** Be open to receive what you want. This means being willing to take the opportunities that present themselves, to act with confidence and not to be discouraged by temporary failures.

The science behind the power of attraction

Although the power of attraction is not yet fully understood by science, several theories attempt to explain how it works. Some of these theories include:

- **Quantum theory:** According to quantum physics, reality is not fixed but exists in a state of infinite potential. Our thoughts and emotions would influence the collapse of this wave function, resulting in the manifestation of a specific reality among infinite possibilities.

- **The placebo effect:** The power of the mind to influence the body is now widely demonstrated.

www.ingramcontent.com/pod-product-compliance
Lightning Source LLC
Chambersburg PA
CBHW032211220526
45472CB00018B/1075